MESHTASTIC®
MADE SIMPLE
HOW TO SET UP AN EMERGENCY RADIO NETWORK FOR ENCRYPTED OFF-GRID COMMUNICATION

MESHTASTIC®
MADE SIMPLE

HOW TO SET UP AN EMERGENCY RADIO NETWORK FOR ENCRYPTED OFF-GRID COMMUNICATION

DAVE HAYES

dhayesMEDIA

© Copyright 2025 – Dave Hayes

All rights reserved. This book is protected by the copyright laws of the United States of America. No portion of this book may be stored electronically, transmitted, copied, reproduced or reprinted for commercial gain or profit without prior written permission from DHayes Media LLC. Permission requests may be emailed to: dave@dhayesmedia.com or sent to the DHayes Media mailing address below. Only the use of short quotations or occasional page copying for personal or group study is allowed without written permission.

DHayes Media LLC, 137 East Elliot Road, #2292, Gilbert, AZ 85299

This book and other DHayes Media titles can be found at: dhayesmedia.com

Available from Amazon.com and other retail outlets.

For more information visit our website at dhayesmedia.com or email us at dave@dhayesmedia.com

ISBN-13: 978-1-966987-99-4
Printed the U.S.A.

DISCLAIMER

The information contained in this book is provided for use by the general public. This book contains non-affiliate products and parts. If you choose to purchase of any of these parts or products, I receive no commission. I've provided product images and links simply for your convenience. The information I've provided is accurate to the best of my knowledge at the time of publishing. **Meshtastic**® is a rapidly evolving field of interest. The protocols are likely to undergo small changes soon after this book is published, but the basic concepts presented will remain relevant as the project grows.

Meshtastic® is a registered trademark of Meshtastic LLC. Meshtastic software components are released under various licenses, see Meshtastic GitHub for details. No warranty is provided – use at your own risk.

ACKNOWLEDGMENTS

I want to thank the development team at Meshtastic® for their assistance in compiling this book, particularly J.M. Casler, who reviewed the manuscript document and provided helpful information.

I must also extend gratitude to my friend Walter Dilley, who contributed his time and expertise by adding essential details and edits to the manuscript's content.

TABLE OF CONTENTS

    Disclaimer 5
    Acknowledgments. 6
    Introduction 9
1   How Does Meshtastic® Work? .13
2   Hardware. 17
3   Software 25
4   Getting Started 31
5   Sending Messages 37
6   Setting Up a Neighborhood Network. 43
7   Hop Limit 47
8   Device Roles. 49
9   Creating and Managing Channels 53
10   Modem Settings 57
11   Privacy and Security 61
12   Antenna Considerations. 65
13   Elevated Locations. 71
14   Remote Node Administration .81
15   Meshtastic and the Internet 83
16   Meshtastic and Amateur Radio. 87

INTRODUCTION

My interest in emergency preparedness began in the 1990s when I worked as a paramedic firefighter in the Pacific Northwest. I became licensed as an amateur radio operator in 2021, and in 2023, published a book titled: *Emergency Preparedness and Off-Grid Communication*. That book includes a short chapter about Meshtastic® but it seemed good to write a book dedicated solely to this subject, due to its growing popularity and its potential for use as an off-grid communication solution. There are many other applications for Meshtastic, and they will be discussed in this book, but my personal focus is on its potential to provide secure, inexpensive communication in the event of a large-scale loss of cellular and internet service.

Imagine being able to communicate without relying on cellular networks, Wi-Fi, or the internet. Imagine a system that allows you to stay connected with others across vast distances, even in remote locations, using only tiny, low-power devices that require no license to operate. Welcome to the world of Meshtastic, an open-source communication project that allows users to create resilient, long-range inexpensive, mesh networks.

When disaster strikes, communication is often the first lifeline to break. Hurricanes, earthquakes, floods, wildfires, and widespread power outages can shut down cellular networks, disrupt the internet, and isolate communities. In these moments, the ability to stay connected with loved ones, coordinate rescue efforts, or access critical information can save lives. This is where Meshtastic steps in—a powerful, affordable, and easy-to-use tool for creating resilient communication networks that operate even when traditional systems fail.

Meshtastic combines cutting-edge LoRa (Long Range) radios with open-source software to build decentralized, long-distance, low-power encrypted mesh networks. These networks allow devices to relay text messages across state lines, providing communication capabilities without relying on cell towers, satellites, or internet infrastructure.

Here are a few of the applications of Meshtastic:

OUTDOOR ADVENTURES | Picture this: you are hiking in a remote national park with no cellular coverage. With Meshtastic devices, your group can stay connected, sharing GPS locations and text messages across miles of rugged terrain. No more worrying about losing track of each other in the wilderness.

EMERGENCY RESPONSE | In disaster situations where infrastructure is damaged, Meshtastic networks can be quickly deployed to coordinate rescue efforts. For example, first responders can use it to share critical information, like survivor locations or resource requirements, without needing cellular towers or the internet.

COMMUNITY NETWORKS | Meshtastic can be used to create resilient community communication systems, whether for neighborhood watch programs, community event coordination, or local projects. Its decentralized nature ensures that the network remains functional even if some nodes go offline.

INTRODUCTION

INTERNET OF THINGS AND AUTOMATION | From monitoring weather stations in remote locations to managing agricultural sensors across vast farmlands, Meshtastic is a powerful tool for Internet of Things (IoT) enthusiasts. Its ability to handle small packets of data efficiently makes it ideal for controlling devices or gathering environmental data.

EDUCATION AND EXPERIMENTATION | Meshtastic is a fantastic platform for learning about wireless communication, networking, and open-source development. Schools, universities, and makerspaces can use it to teach students about real-world technology.

TACTICAL AND FIELD OPERATIONS | In professional fields like search and rescue or military operations, Meshtastic can be used for secure, short-range communication over large areas without requiring expensive equipment. Meshtastic can be used together with the popular ATAK mobile app.

Traditional email and text messaging relies on networks others have built, but mesh networks are created and maintained by their users. Meshtastic enables you to build your own communication network. If that sounds intimidating, don't worry. It's not difficult. Whether you're a seasoned tinkerer or a beginner, this book will guide you step by step as you build and deploy your own network, regardless of your technical background.

1: HOW DOES MESHTASTIC WORK?

Most people are familiar with the dynamics of two-way radio communication, where two people operating handheld radios send voice messages to each other. To communicate successfully, both radios must be set to the same frequency, both operators must be within range, and both operators must be listening or available at the same time. Mesh networking uses a completely different dynamic.

A mesh network is a system where two or more radios send messages back and forth continually. Once the radios are powered on, they begin transmitting and receiving automatically. Rather than two radios communicating to each other, a mesh network may consist of dozens or even hundreds of radios communicating simultaneously. Each device communicates directly with other devices that are within range, and can forward messages to other nodes within the network.

A *radio* is a *device* that can be connected to other devices to form a *mesh network*. Each device in a network is called a *node*. In this book, the terms *radio*, *device*, and *node* are used interchangeably.

HOW DOES MESHTASTIC WORK? | Imagine a hiker who sends a message from a remote mountaintop to his wife. If the hiker is too far from his wife for direct communication, but other hikers with connected nodes are between them, the message can hop through other hikers' devices, and relay the message to the wife.

Meshtastic® uses LoRa radios, which are designed to send data over long distances at low power. Users connect a phone or table to their device via Bluetooth or Wi-Fi. The phone uses a mobile app interface that allows the user to send text messages to the radio, which transmits a message that can be received by any radio within range. The Meshtastic software enables nodes to automatically discover each other when they are within radio range. The mesh protocol ensures that messages can "hop" from one device to another to reach the intended destination, even if the destination is outside the direct communication range of the originating device.

Most radios require users to be listening at the time a message is sent because most two way radios do not allow messages to be stored. Meshtastic works with a smart phone or tablet, which stores messages as they are received. One brilliant feature is the fact that if your radio is connected to the mesh network, it will save messages, even if your phone is not within range of the device. When it is convenient, you can reconnect your phone to the device, and messages that were received while you were away will be downloaded to your phone.

Unlike traditional networks that rely on a central server or infrastructure such as cellular towers and Wi-Fi routers, mesh networks in general operate without a central point of failure. If one device goes offline, the network can continue functioning as long as there are other devices to interact with.

Mesh networks can be scaled according to need. They can be expanded by adding more radios. The effective range of a network increases as more devices are added. However, very large networks develop performance problems, which will be addressed in a later chapter.

1: HOW DOES MESHTASTIC WORK?

Mesh networking offers resilience. The more nodes there are in a network, the more potential paths messages can take. If one path fails, a message can find another route, making the network more reliable and resistant to interference.

2: HARDWARE

A step-by-step guide will be provided to help you build your own mesh network. The process begins with selecting hardware. Meshtastic® utilizes low power radios built on small printed circuit boards. Because the radios use little power, battery equipped units can operate for long periods on a single charge. Battery life is extended by software that periodically shuts down non-essential functions. This "sleep" mode allows Meshtastic devices to run on a single battery charge for days or even weeks.

If you're looking for a device that requires no assembly, several options are available. Preassembled Meshtastic radios are sold by several manufacturers. Rokland sells the WisMesh pocket, a small, battery powered device with a 3D printed case. They also offer the WisMesh Tap; a weatherproof radio with a touchscreen interface. Both models require no assembly by the user.

ROKLAND WISMESH POCKET   ROKLAND WISMESH TAP

Continuing with devices requiring no assembly, LILYGO offers the T-Echo, an inexpensive battery powered unit (below left).

If you want an integrated keyboard, the LILYGO T-Deck (below right) is a popular choice that resembles a Blackberry phone. It can be purchased with or without a case from several retail outlets, including Amazon.

Etsy has many vendors that sell preassembled Meshtastic devices in a range of prices and specifications. LoRa devices can also be built by the user, and it's not as complicated as you might imagine.

As I mentioned, Meshtastic radios are integrated into small printed circuit boards. Each board has a USB port for power, a small radio transmitter, an antenna connector, and a connector for a battery. Some boards have GPS modules or connectors for solar panels. The required software is downloaded onto the board through the USB port. That process is covered in the next chapter.

Meshtastic hardware suitable for do-it-yourself builders include devices like the LILYGO T-Beam, T-Beam Supreme, several Heltec development boards, and modules made by RAK Wireless. A list of supported devices is available on the Meshtastic.org website. To find them navigate to Hardware > Supported Hardware > Devices.

2: HARDWARE

Providing a source of power should be a consideration when choosing your device. The LILYGO T-Beam shown below has an optional holder for a rechargeable 18650 style Lithium-ion battery.

Before choosing a device, consider your specific needs. If you do not need GPS, consider a board that doesn't have a GPS module. It will use less power.

Some boards like the Heltec LoRa 32 V3 use more power when compared to other boards and require frequent recharging. A board with a display screen will use more power than a similar board without one. Strictly speaking, display screens are not necessary since Meshtastic relies on a phone and mobile app for the user interface.

In December 2024, Heltec made available their Mesh Node T114 V2. It's a redesign of an earlier model with major improvements. The T114 is built on a LoRa board that uses a fraction of the power the Heltec ESP32 V3 uses. I connected one to a 950 mAh battery and it lasted four days on a single charge with occasional use. This unit sells for around $35. For a few dollars more, it can be purchased with a plastic case that encloses all its components. The case has room to accommodate a 950 mAh battery.

Some LoRa boards incorporate modules for barometric pressure, temperature, and other environmental conditions. If your use case requires this data, consider buying one of these boards. But the more data a unit acquires or is needed more frequently, the more it will transmit, and that has a downside: it drains the battery quicker.

Some applications will benefit from a simple, energy efficient board without the bells and whistles. My favorite is the Rak 4631 Wisblock. It has no display screen, it's energy efficient by design, and has the option to add a solar panel for charging the battery. (The 4630 is a smaller and less expensive version.)

RAK 4631 WISBLOCK

You could operate a Meshtastic board by simply connecting the printed circuit board to a USB power supply (or battery) and an antenna, but this doesn't allow for portability. To protect the device you'll need an enclosure of some kind. An enclosure shields the electronics from dust, moisture, and impact, and allows for a permanent battery to be installed. It also keeps the antenna from accidentally disconnecting and to maintain its vertical orientation.

If you have a 3D printer, you can print a case for your device. Print files for most LoRa boards can be found on sites that aggregate 3D print files. I designed a 3D printed case for the Heltec LoRa 32 V3 board. The body of the case has two compartments. The lower compartment accommodates a rechargeable 3,000 mAh battery. The upper compartment houses the Heltec board and antenna cable. Once the components are placed inside, a cover is installed that has a window for the display screen, an opening for the USB port, and "buttons" on the face that allow the user to operate the power and reset buttons on the board. This device will run for 2.5 days with light use on a single charge.

2: HARDWARE

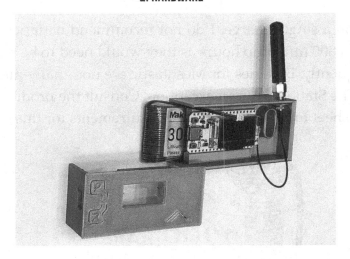

If you 3D print your own Meshtastic case, I recommend using PETG filament. It will survive hot climates better than PLA or other materials. I also recommend Creality brand filament. I've tried other brands and generally found the finished product to be more brittle, which causes 3D printed objects to be easily broken.

BATTERIES | Unlike a handheld radio, a Meshtastic device does not need to be held in the hand for it to send messages. Messages are (generally) composed on your phone using an app. As long as the LoRa radio is within Bluetooth range of your phone, you can send messages from your phone to the radio. Thus, you could place a Meshtastic device on your desk, plug it in to a power cord that provides household current, and it would not need a battery. It wouldn't even need a case, though its aesthetic value would suffer. Adding a battery makes it portable, and a case to hold the components together can provide a pleasing form.

Most LoRa boards have JST connectors that accept rechargeable batteries. Rechargeable batteries are rated as to how much voltage they provide and how many ampere hours of service can be expected on a single charge. I prefer batteries rated between 1,000 and 3,000 milliamp hours (mAh). One could connect a battery rated at 10,000 mAh if one needed. This would provide weeks of use on a single charge. I do not recommend batteries rated at less than 800 milliamp hours as they would need to be recharged too frequently. Batteries for Meshtastic are nominally rated at 3.7 volts. (The Station G2 is an exception. Consult the product's information sheet for details on power requirements for this device.)

Pay close attention to the connectors provided with any battery you purchase. The size and type of connector varies widely depending on manufacturer. Most LoRa boards accept a 1.25 mm JST battery connector. This information should be included in the battery description. The exception is the Rak Wisblock, which has a 2.0 mm JST connector for the battery and 1.25 mm connector for the solar panel.

2: HARDWARE

Because I build Meshtastic units on both the Heltec LoRa 32 V3 and the Rak Wisblock boards, I've opted to buy batteries in bulk with the 1.25 mm JST connector. When I use this battery for a Rak board, I cut off the 1.25 mm connector and splice on a 2.0 mm connector. The 2.0 mm JST connectors can be purchased online through Amazon.

And now a word of caution: Be advised that many aftermarket JST connectors have reversed polarity. I found this out by burning up a couple of Rak boards. When a JST connector has reversed polarity, I splice the red wire to the black wire coming from the battery and the red wire from the battery is attached to the black wire from the JST connector. The easy way to know you're connecting a battery correctly is to remember that the positive pole is on the inboard side of both the solar and battery sockets on the Rak board (The positive and negative poles are marked on the board.). On most batteries, the red wire is positive, so it should lead to the inboard (positive) side of the JST connector.

3: SOFTWARE

This chapter explains step-by-step how to install the software needed for Meshtastic® use. Begin by installing the latest version of the Meshtastic app on your phone or tablet. The app is available in both Android and iOS app stores. Alternatively, it can be found in the Meshtastic GitHub repository and sideloaded onto your device.

If operational security is a concern, consider the following: The phone you will use does not need to have a SIM card installed. I have several older Android phones without SIM cards that I use for connecting to Meshtastic nodes. I turn on the phone's Wi-Fi and connect to my home internet just long enough to download and install the app. Once installed, I turn the Wi-Fi off and it is not needed again. If you don't want to log in on the phone as a user, I recommend downloading the Meshtastic apk file from Github. Once it has been downloaded, install it. Depending on your device security settings, you may need to manually approve the installation.

If you prefer to use a computer as an interface, you can manage your device from a web browser at this address: https://client.meshtastic.org/. If your device has the ESP32 chip, download and install the CP210X USB to UART bridge driver from the Meshtastic website. Navigate to the ESP32 drivers tab and download the driver for your operating system.

Another alternative is the MeshSense program by Affirmatech, which has some unique features that make it worth considering, including a graphic display of devices that are connected directly by radio signal. To use MeshSense, download the program and install it, then connect your computer to your device via IP, Wi-Fi or Bluetooth.

If you purchase a preassembled device, it should come with the Meshtastic firmware installed. In some cases, the installed firmware may not be compatible with the version of the mobile app you are using. In this case, you will see an error message on the app when trying to pair the phone to the device. The message will indicate that the firmware on the device must be updated.

If your device does not have firmware, or if the installed version needs to be updated, or if you decide to build your own device, you can install the firmware by navigating to the Meshtastic web flasher page on a Chrome or Edge based browser.

The workflow for installing firmware is slightly different depending on the model of board one is using. I would like to note here that the words "flash" and "download" are used interchangeably when referring to the installation of Meshtastic firmware on an ESP32 Lora board since the program is downloaded to the device and not to a computer. Most Heltec, LILYGO and other boards built upon the ESP32 platform can be flashed by following these steps:

To install the firmware, make sure a suitable antenna is connected to the device. Meshtastic radios transmit as soon as they are powered on, and the transmitter may be damaged if an antenna is not connected. Preassembled devices usually come with an antenna that can be screwed on to an antenna port. If you're building your own device, you'll need to attach an antenna to the LoRa connector of the board. Some boards have connectors

3: SOFTWARE

for both Bluetooth (BLE) and LoRa antennas. Most manufacturers provide the necessary antennas. It isn't required at this time to connect a Bluetooth antenna, but an antenna should be connected to the LoRa socket before it is powered on. (The LoRa socket should be labeled as shown in the example below.)

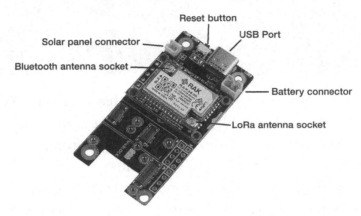

Next, connect the device to a computer with a **data capable** USB cable. In most cases, the board will require a USB type C cable. Once the cable is connected to the device and your computer, the device will power on. Note: not all USB cables are able to transfer data. Some are only designed for charging the battery of a device. Most phones are supplied with data cables, so try the one that came with your phone, if the connector is compatible with the board's USB port. Boards built by Rak Wireless usually ship with a data cable included. Most aftermarket USB cables sold by Anker offer data transfer capability.

Once the device is connected to a computer, and you've navigated to the Meshtastic web flasher page, from the drop-down menu on the left, select the model of the device you are using. (For this demonstration, I will select the Heltec V3 board.) Next, select the firmware version from the drop-down menu to the right. I recommend the latest stable (beta) version. The latest version will be at the top of the list, with the older ones being found near the bottom. Next, press the "flash" button.

MESHTASTIC® MADE SIMPLE

A new window will open that shows the features of the version you selected. At the bottom of the window, press the "Continue" button. A new window will open. Follow the prompts that are provided. You can use the default baud rate or select a higher one for faster flashing. Toggle the button to the right to confirm "Full erase and install." With ESP32 boards, a bar will appear at the bottom of the window labeled "Erase Flash and Install." Press the bar to continue.

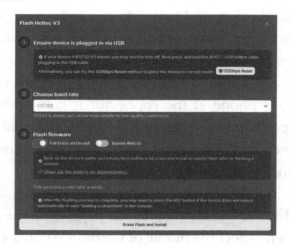

A popup window may appear asking you to confirm the serial port to be used. Select the com port the device is connected to. If multiple ports are available and you're not sure which one to choose, write down the number of each port displayed, i.e., COM6, COM3, etc. Disconnect the device you're flashing. Try flashing

3: SOFTWARE

again and note which port is missing from the list. The one that is missing is one you need to connect to. Re-connect the device to be flashed and select that port.

The device will begin flashing. Once completed, it should automatically reboot. It is now ready to use.

RAK WIRELESS | On the final page in the flashing process, Rak boards (and other devices built on the nRF52 se-ries platform) will show a button labeled "Enter (UF2) DFU mode." Press this button to confirm. Rak boards are flashed by dragging and dropping a UF2 file to a folder that displays on your computer. To display the folder, after the device is connected to your computer via USB cable, press the reset button twice quickly. (The reset button is located beside the USB port.) The Rak board will appear as an external drive on your computer. Select this drive and a folder will open. Download the UF2 file to your computer. Next, drag and drop the UF2 file to the Rak folder. When the firmware has been flashed, the Rak folder will close and the device will reboot. It is now ready for use.

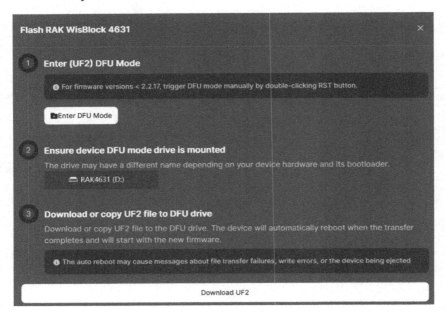

4: GETTING STARTED

Once your phone or tablet has the Meshtastic® app installed and your device has been flashed, you can start to set up your network. Your first mesh network can be built in just a few minutes using two devices. This chapter will explain how to do it.

After connecting an antenna, power your device on. To prove that Meshtastic doesn't require internet or cellular service, put your phone in airplane mode. This will disable all external data services. Now, turn on just the Bluetooth function. Next, open the Meshtastic app on your phone or tablet. The main page for the app has five tabbed pages. The page farthest to the right has a gear icon. This is the connection page. (The following images are from the Android version of the mobile app. The iOS version is similar, but not the same.)

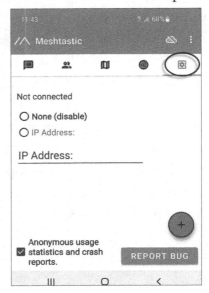

From this page, press the plus (+) button in the lower part of the page and a window should open and display available Bluetooth devices.

Select the device you wish to connect to. **Tip:** the first time it is used, the device will have a name like "Meshtastic 852b." The designator "852b" will appear in small text on the display screen of the device if it has a screen. This will help you know you're pairing to the correct device if there is more than one available in the immediate area.

On your phone, select the device to be paired. If the device has a display screen, a six-digit PIN code will appear on the screen. If the device does not have a display screen, use the default PIN code: 123456. Enter this PIN code into the pairing dialog box on your phone.

If pairing is successful, the app will display a cloud icon in the upper right corner with a check mark. If pairing is unsuccessful the cloud icon will have a slash through it. If you are unsuccessful on the first attempt, try again. It may take several attempts to pair successfully.

The cloud icon in the top right corner indicates whether you are connected to a device.

4: GETTING STARTED

This feature currently has three states: a cloud with a slash through it indicates no device is connected to the app. A cloud with a check mark means a device is connected to the application. A cloud with an up arrow indicates a device is connected, but it is currently sleeping or out of range.

After successful pairing, the first thing to do is set your region from the dropdown list on the right side of the connection page. Anytime a change is made on the app, the new settings will be sent to the device and stored. Then, the device will automatically reboot. In most cases, the device will automatically re-pair with your phone, though you may need to re-enter a PIN code.

The next step is to give your node a long and short name. To do this, navigate to the Radio Configuration page. This is done by tapping the three dots in the upper right corner of the app. From this page select "User." In the dialog box, enter a long name that is less than 40 characters and a short name of no more than four characters. If you intend to use the device under an amateur radio license, toggle the switch on. Be advised that this option disables the use of encryption. When you are finished, press the "Send" button. The device will automatically reboot.

The tab that is second from the left on the app has an icon of two people. This page lists connected nodes. The device your phone is connected to will appear at the top of the node list. The default settings will display connected nodes below it in chronological order of when they were last detected. This list can be sorted alphabetically from the drop-down menu. The search bar allows you to find a particular node by name from the list. Tapping on a node's short name will display a host of features including device metrics for any connected node.

After pairing, it's possible that the app will show a nearby node owned by a neighbor. If so, you have established a network consisting of two nodes. If not, you can add a second device of your own by following the steps above with a second phone or tablet and a second radio. Keep in mind that a phone can only be connected to one node at a time, and a node can only be connected to one phone at a time. If you have two phones or tablets, and two radios, try pairing a phone or tablet to each node. Once both devices are paired, you have a simple mesh network.

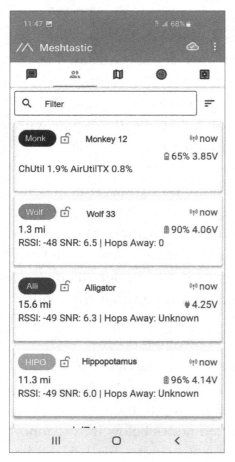

After the app has had time to analyze the connected nodes, data will appear with each node regarding connection status. The RSSI value is the Received Signal Strength Indicator. A node that your device is connected to by direct radio frequency will show an RSSI and Signal to Noise Ratio (SNR). Any node connected by intermediate nodes will instead display a "hop" value indicating how many nodes are between it and your device. Battery status is usually displayed for connected nodes as well as weather conditions for nodes enabled with those features. If your device is set to report its position, the distance between it and connected nodes will be displayed in either kilometers or miles. To change from metric to imperial, go to the "Display" tab in the Radio Configuration page and adjust this to your preference.

4: GETTING STARTED

GPS LOCATION | On the middle tab of the app, you will find a map page that displays the locations and short names of connected nodes that have enabled position reporting. Not all Meshtastic devices are enabled with GPS modules. If your device has GPS, by default, its location will be displayed on a map and can be seen by anyone in range who uses the default LongFast channel. If your device does not have a GPS module, but you allow the app to use your phone's GPS location, it will display the location of the phone on the map.

If security is a concern, you can disable GPS in the Radio Configuration page under the "Position" tab. When position is disabled, a node in the default Client role will appear in the list of connected devices, but it will not appear on the map. (Device roles are discussed in another chapter.) The Position section has an option to adjust the accuracy of GPS location, if you want to give a vague, rather than a precise location. This section also allows you to set a fixed position using latitude, longitude, and elevation. Setting a fixed position may be useful when laying out the location of nodes for a network. (If security is a concern, you can spoof your device's location to anywhere you want.)

5: SENDING MESSAGES

Once two or more devices are connected, you can begin sending messages between them. Encrypted text messages can be sent either directly to a particular node, or broadcast to a channel.

To send a direct message to a particular node, in the list of devices, tap on the colored oval that displays the four character name of the node you wish to communicate with. A drop-down menu will appear. Select "Direct Message."

A new window will open. Near the bottom of this page, there will be a text box. Compose your message and press the arrow to the right of the text box to send it. When sending a direct message, an icon will appear beside it indicating the status of the message. The five icons are listed below with their meanings.

- A slash through a cloud means the message was not sent. Try composing and sending it again. Messages can be copied and pasted from the list of messages into the text field using the menu at the top of the page. There is also an option to delete messages.

- An up arrow inside a cloud indicates the message is queued for transmission.

- An empty cloud indicates the message is in transit but has not reached another node.

- A check mark in the cloud means the message has been received by at least one other node.

- A check mark inside a human head appears when the message has been received by the intended recipient.

CHANNEL MESSAGES | A message can be sent to a group by using the channel feature. By default, the Meshtastic® app sets up a publicly accessible channel for the network. The default channel is named LongFast. Think of a channel as a community bulletin board where messages intended for general dissemination are posted. Message threads are found on the message page, which is the tab farthest to the left. Navigate to this page and tap on the LongFast

5: SENDING MESSAGES

channel. By default, anyone within radio range who is connected to the network can post a message in this channel.

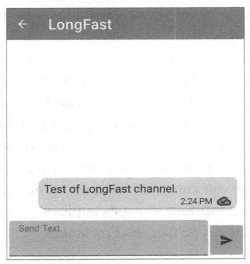

In the text box, near the bottom, compose and send a message. If you have a second device connected, it should receive the message and display it in the LongFast channel on the messages tab. If you have notifications enabled, the phone paired to the other device should receive a notification.

Messages sent in channels utilize a confirmation method that is different from direct messages. Because a direct message is sent to an individual device, it can be determined whether the device received it, and that fact can be confirmed. A message sent to a channel is not intended for a particular device. Instead, these messages are received by any node that is connected and using that channel. It cannot be determined if an intended device received a channel message since it is intended for multiple users. Thus, when a message is sent to a channel, and is received and relayed by at least one other device, a check mark in a cloud will appear beside the message. You can reply to this message by following the steps above. An empty cloud means the message is in transit. A cloud with a slash through it means the message failed to send.

In the messages tab, as direct messages and posts to channels are sent and received, they will appear as individual threads on this page, like how your phone organizes text messages from people in your contact list.

Now that you understand the basics of Meshtastic, you're free to experiment. You might leave one device at home and take the other for a walk around your neighborhood. Send a message at each intersection with the names of the streets. When you return home, see how many messages were transmitted successfully. The next time a group of friends travel somewhere, give each person a node and see if you can communicate while on the road. On your next camping trip, take along a couple of radios and see how far from the campsite you can send messages.

ISSUES TO CONSIDER | Although Meshtastic radios will automatically acquire signals from nearby devices and add them to the network, it is not an instantaneous process. Some nodes will be detected within minutes. Others may take hours to acquire. It may take 24 hours to compile a complete list of nodes in your area.

Although the Meshtastic app provides a default channel for general use, up to eight channels can be set up. Channel settings will be discussed in another chapter.

TROUBLESHOOTING | If a Meshtastic device fails to pair, check to make sure it is powered on. If it is on, press the reset button to make sure it is not in "sleep" mode.

It's not unusual for a device to fail to pair with a phone on the first attempt. Try pairing several times before going further in the troubleshooting process.

If a radio is not displayed in the list of devices available for pairing, check to make sure you are pairing to the correct device ID. If the unit has a display screen, the device ID will be displayed

5: SENDING MESSAGES

on the screen. Scroll through the display screens until it is visible, then make note of it.

If a device fails to pair, it could be that the Meshtastic firmware is missing or corrupted. The solution is to reflash the firmware, wait for the device to reboot, and try pairing again.

In some cases, pairing problems can be resolved by deleting (forgetting) previously paired devices from your phone's Bluetooth settings and starting with an empty list.

6: SETTING UP A NEIGHBORHOOD NETWORK

Creating a Meshtastic® network for your neighborhood is an excellent way to build a resilient and independent communication system for staying connected, sharing updates, or coordinating social activities or communicating during emergencies. This chapter will guide you through the entire process, from planning and hardware selection to deployment and testing.

When planning for an expanded network, one issue to consider is the range of a Meshtastic device. A typical voice handheld radio will produce a signal of roughly 5 watts. At ground level, because radio signals are absorbed by vegetation and buildings, in a suburban setting, the effective range of two handheld radios is about one mile. If the same radios were used from points on two mountaintops, the range would be dozens of miles, because there would be no objects in the way to absorb the signal.

LoRa radios produce a signal of roughly 150 milliwatts (1/7 of a watt.). This weak signal is easily absorbed by buildings and trees, making the range of Meshtastic devices around ¼ mile in a suburban setting at ground level. When mounted above the roofline of a house, the range can be greatly extended. I have a roof mounted node at my single-story home in Phoenix, Arizona. I live in a typical suburban neighborhood and regularly receive signals directly from nodes in Tucson 95 miles away. The old amateur radio adage "height is might" is true for Meshtastic. A node deployed at rooftop height, can serve as a central hub and relay traffic to other nodes nearby that are used at ground level.

Since the network must always be available, especially in an emergency, the nodes that will serve as the backbone for the network require a constant source of power that cannot be interrupted. Because Meshtastic radios operate on extremely low power, a 2-watt solar panel that costs a couple dollars and an appropriately sized rechargeable battery is ideal for this purpose. Most low power solar panels come with USB C cords, which makes them easy to connect to a Meshtastic device.

A second consideration is weatherproofing. A backbone node could be deployed in an attic space, which would protect it from rain and snow, but it would be positioned just below the roofline and its range would be less than one deployed above the roofline. I would not rule this option out, but a weatherproof node deployed outside would be a better option.

ROKLAND WISHMESH TAP

The WisMesh Tap sold by Rokland is worth considering for the application. It's weatherproof and can be deployed outdoors. It has a USB C port that can accommodate a solar panel power cord. The WisMesh Tap has recessed threaded brass inserts on the back to accommodate mounting screws. It would just be a matter of engineering a suitable mounting device for the radio and solar panel. More options for weatherproof devices are covered in the chapter Elevated Locations.

The strategy for building a neighborhood network is to deploy at least one, and perhaps several nodes (depending on the size of

6: SETTING UP A NEIGHBORHOOD NETWORK

the area to be covered) at rooftop level or higher. A tower or tall building is a good place to deploy a weatherproof node. Once you have a node deployed in an elevated location, you can use it to connect radios in your home, on your person, or in your vehicle. The elevated node acts as a kind of repeater, relaying messages between devices in the home and those deployed in the area.

In addition to elevation, antenna gain is a factor that affects range. The standard stubby antenna supplied with most Meshtastic boards is not well suited for the 900 MHz band, and they tend to be inefficient. Stock antennas also provide little in the way of gain. The design of an antenna determines the amount of signal it sends and receives in a particular direction. High gain antennas focus the signal in a specific path, giving them the ability to send the signal farther than lower gain antennas. A 5dBi antenna is preferred over the standard antenna supplied by most manufacturers. Even better is a 12dBi antenna, especially if your goal is setting up a regional network where distance is critical. More information on antennas is provided in the chapter Antenna Considerations.

Once you have a node deployed in an elevated location and several are connected at ground level, you can begin testing the range and reliability of the network. Send direct messages to all other connected nodes including the ones in elevated locations. Note whether it takes two, five, or ten attempts to successfully transmit a message to a particular device. If it takes several attempts, consider whether that node would benefit from a higher gain antenna or whether an additional elevated node nearby is needed. Send messages on all the channels you've created and test the ability of other nodes to receive them and reply.

Test the ability of stationary devices to send and receive messages, but also, test the reliability of nodes used inside vehicles. An automobile chassis can impair the ability of radios to receive signals. It's worth finding out early in the testing process if mobile nodes can be relied upon.

As you're experimenting, note whether your network includes nodes in distant communities. Consider contacting the owner of a distant node by direct message and ask if they're willing to participate in your testing and if they have similar goals in mind.

7: HOP LIMIT

Most of the default settings found in the Meshtastic® app can be left as they are. But one setting that has been the subject of considerable debate is the **Hop Limit**, which is found in the LoRa settings tab.

Meshtastic nodes pass data from one device to another. Each time a device rebroadcasts a data packet it is called a "hop." Users can set a limit to the number of hops allowed for data sent by their device. This is called the "hop limit." Here's how it works:

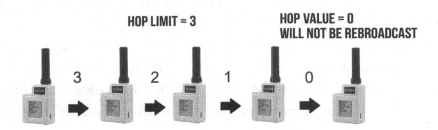

As a data packet travels from one device to another through the mesh, each time it is rebroadcast, one "hop" is subtracted from the value set as the "hop limit." For example, if a device has a hop limit of 3, when a data packet is sent, the next device will remove one "hop" from the current value, making the current value 2. It will then rebroadcast the packet. The next device does the same thing, as do the other devices, until the current value becomes zero. A device that receives a data packet with a value of zero will not rebroadcast it.

The hop limit can be changed by navigating to the Radio Configuration page and then the LoRa tab and adjusting the Hop Limit value. However, Meshtastic developers recommend leaving the hop limit set to 3.

The need for a hop limit arises from the tendency of mesh networks to generate redundant data packets that clog up the network, and reduce efficiency. Setting a reasonable hop limit prevents this problem, which is particularly troublesome with large networks.

In 2024, the metropolitan Phoenix area exploded with Meshtastic activity. In just a few months, hundreds of new nodes came online. Meshtastic uses a data routing protocol called "managed flood routing" to send data packets through the network. The pathways taken by a particular data packet may not be the most efficient route available. As a network grows, thousands of data packets jam up the small amount of bandwidth that is available. When users increase the hop limit of their devices, it aggravates this problem. By fall of 2024, in the Phoenix area, it became nearly impossible to send a direct message successfully. To minimize congestion on busy networks, the best practice is to leave the hop limit at 3. In urban areas, alternate frequency slots may be used to build subnetworks that handle increased traffic. The use of alternate frequencies will be discussed in another chapter.

8: DEVICE ROLES

The Meshtastic® app allows devices to be assigned various roles depending on how the user intends to deploy them. A mountaintop node, for example, has different performance requirements than one used in the home. Roles can be assigned that are appropriate for each use case. This chapter lists the device roles that are available and explains why you might choose a role for a particular node.

Client: This is the default role and it is appropriate for most cases. This is the preferred role when a node is connected to the app and used for messaging other devices on the network. (Despite the apparent baggage of the term *Client* in some technological contexts, Clients in Meshtastic do actually repeat/route messages. Unfortunately, this has caused some confusion, leading some individuals to incorrectly choose the Router role.)

Client Mute: A device assigned this role does not forward packets from other devices on the network. This role is for situations where a device needs to send messages on the network without assisting in packet routing. For example, if you have a rooftop node that connects to a neighborhood network and it is out of Bluetooth range from inside the home, you may use a second device in your home to connect to the rooftop node. In this case, the rooftop node should be set to **Client**, and the node in the home should be set to **Client Mute**

Client Hidden: This role is assigned when security and privacy are top priorities. A device set to this role only broadcasts the minimum number of data packets required. It is not visible in the list of nodes and cannot be seen on the map. Use this role for stealth deployment or to reduce airtime and power consumption.

Tracker: A device assigned this role broadcasts GPS position packets as priority. For use in tracking the location of individuals or assets, especially in scenarios where timely and efficient location updates are critical.

Lost and Found: This role allows a device to broadcast location as a message to the default channel regularly to assist with device recovery when a node is lost.

Sensor: This role broadcasts telemetry packets as priority. For use when gathering environmental or other sensor data is crucial, with efficient power usage and frequent updates.

TAK: This role is optimized for ATAK system communication. It reduces routine broadcasts. Select this role when integrating with ATAK systems (via the Meshtastic ATAK Plugin) for communication in tactical or coordinated operations.

TAK Tracker: This role enables automatic TAK PLI broadcasts and reduces routine broadcasts. It's a standalone PLI integration with ATAK systems for communication in tactical or coordinated operations.

8: DEVICE ROLES

Repeater: This role is assigned to nodes that pass traffic on the network but are not connected to a user for messaging. A device set to **Repeater** is not visible in the nodes list or on the map. These devices are typically deployed in strategic, elevated locations where they expand the network's geographic coverage.

Router: This role is similar to **Repeater**, but these devices are visible in the nodes list and on the map. Router devices are typically deployed in strategic, elevated locations where they maximize the network's geographic coverage.

Roles are assigned from the Radio Configuration page under the "Device" tab. The first option listed on this page is "Role." Tap the arrow to reveal a dropdown menu with the roles that can be assigned. As a rule, **Client** is the role most suitable for app-connected nodes used for sending messages. **Client Mute** is suitable for nodes that connect to a nearby device deployed in an elevated location. Nodes deployed in elevated locations, i.e., rooftops and towers should likewise be set to **Client**. The **Router** and **Repeater** roles should be reserved for devices deployed on mountaintops where data is passed between cities or counties. When in doubt, **Client** is the safest choice.

9: CREATING AND MANAGING CHANNELS

Before beginning this discussion, it will help to make a distinction about the hierarchy of Meshtastic® channels. The Meshtastic community refers to all channels, whether the primary or secondary ones as **channels**. However, the primary channel has special features that other channels do not. For example, the primary channel, which occupies the 0 position in the list on the channels page, determines the modem preset for all other channels. It might be better to think of the primary, or 0 channel as the main channel, and all others (1-7) as subchannels. Each subchannel uses the modem preset of the main channel but has its own name and may have a unique encryption key.

The creation of specialized subchannels allows Meshtastic users to segregate communication within their network, enabling multiple groups of users to operate independently on the same hardware. By creating different subchannels, you can create private message boards, enhance security, or optimize network performance for specific use cases. Rather than thinking of a subchannel as a frequency, think of it as a dedicated chat room.

You might, for example, create a neighborhood subchannel that handles traffic of interest to anyone in your neighborhood. For members of your immediate family, you might create a family subchannel. And for communication between members of your church, you might have a dedicated church subchannel.

Meshtastic subchannels segregate messages into user-defined groups. Each subchannel is defined by a unique name, which can be whatever you want. An optional encryption key can be created for each subchannel for secure communication.

Here are step-by-step instructions for creating a new subchannel:

- Open the Meshtastic app.

- Pair your device via Bluetooth and connect to a node.

- Press the three-dot button in the upper right corner of the app to open the Radio Configuration page.

- Go to the Channels tab. Channels are numbered 0 through 7, with 0 being the primary channel and others being subchannels.

- Tapping the X on a channel will delete it. Tapping the plus (+) will open a dialog box where you can create a new channel.

- In the dialog box, give your new channel a name.

- A new encryption key will be created. You can generate a replacement encryption key for a channel by tapping the circular arrow in the PSK box.

- Press the Save button.

- To add another channel, follow the steps above.

- When you are finished, press the Send button. The device will automatically reboot.

9: CREATING AND MANAGING CHANNELS

Once a new channel has been created, it can be shared with other nodes. For users to communicate on a channel, they must have the identical channel settings. Sharing the channel settings is done on the channels tab, which is the second from the right on the mobile app.

On this page, you will see a list of channels, with the primary channel being found at the top. Additional channels are found below it. Beneath the channels list is a QR code that can be scanned by those with whom you want to share channel information. Below the QR code is a link that can be sent via text or email to those you wish to invite to the channel. Below the link is a box where you can change the modem preset for the channels. Be advised: Changing the modem preset for one channel changes the preset for all channels. Buttons at the bottom of the page allow you to reset the channel or scan a QR code.

Notice that there are boxes to the right of each channel name. As you tap a box, a checkmark will appear in the box indicating that the channel has been selected. Tap the box again to deselect it. As you select and deselect channels, the config-

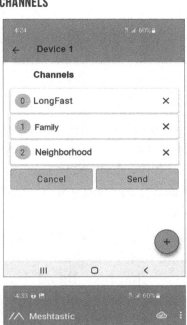

uration of the QR code changes. To share a single channel with someone, select only that channel from the list and allow them to scan the resulting QR code. Alternatively, the QR code can be captured as a screenshot and emailed or sent via text message. A third option is to send the link generated in the box below the QR code. To share all your channels with someone, select all channels and have them scan the QR code or send them the link. It's a good practice to send a test message on a new channel to verify that it is working properly. As a matter of habit, when I add a new node to be deployed on my network, I scan the QR code of an existing node to the new device so it can communicate with other nodes on all channels. Be advised, when scanning a QR code from the app, there are options to add or replace channels. The selection to "add" or "replace" is made by tapping one of these choices on the app. Additionally, you must check or uncheck the appropriate boxes beside each channel you intend to add.

10: MODEM SETTINGS

LoRa radios that are used to build mesh networks offer many modem settings to balance communication speed and range. One lesson we've learned in the Phoenix area is that relying on the default LongFast modem setting does not provide the needed data throughput for networks with more than a few dozen users. Meshtastic® offers a set of eight predefined modem "Presets" based on the Semtech LoRa calculator. Each preset is tailored for a specific scenario. Understanding these presets is crucial for optimizing your network's performance. Modem presets provide choices that favor either data rate (speed) or link budget (range). They are arranged below from fastest speed with shortest range to slowest speed with longest range:
- ShortTurbo
- ShortFast
- ShortSlow
- MediumFast
- MediumSlow
- LongFast (Default)
- LongModerate
- LongSlow

Each modem preset is defined by three key parameters:

Bandwidth: The width of the frequency spectrum used. Presets with faster speeds utilize greater bandwidth.

Spreading Factor: Determines the number of chirps per symbol, affecting data rate and range.

Coding Rate: The ratio of data bits to total bits transmitted, which influences error correction capabilities.

Listed below are the most notable advantages and disadvantages of each modem preset:

Short Range / Turbo:

Advantages: High data rate with moderate range, balancing speed and coverage.

Disadvantages: Reduced range compared to slower presets.

Short Range / Fast:

Advantages: High data rate with moderate range, balancing speed and coverage.

Disadvantages: Reduced range compared to slower presets.

Short range / Slow:

Advantages: Improved range over Short Range / Fast while maintaining reasonable data rates.

Disadvantages: Slightly slower data transmission.

Medium Range / Fast:

Advantages: Offers balanced performance for medium-range communications with acceptable data rates.

Disadvantages: Not optimized for either extreme of speed or range.

10: MODEM SETTINGS

Medium Range / Slow:

Advantages: Extended range suitable for larger areas, with moderate data rates.

Disadvantages: Slower data transmission compared to faster presets.

Long Range / Fast (Default):

Advantages: Offers a strong mix of speed and range, suitable for most users.

Disadvantages: May not be optimal for scenarios requiring either maximum speed or maximum range.

Long Range / Moderate:

Advantages: Enhanced range with reliable data transmission, suitable for challenging environments.

Disadvantages: Lower data rates, leading to longer transmission times.

Long Range / Slow:

Advantages: Significant range extension, ideal for long-distance communication.

Disadvantages: Very slow data rates, not suitable for time-sensitive applications.

Here are additional considerations when choosing a preset:

Network Congestion: In networks with high device density or frequent messaging, faster presets can help reduce congestion by decreasing the airtime used per device.

Environmental Factors: Obstacles, terrain, and urban environments can affect signal propagation. Selecting a preset with a higher link budget (offering longer range) can help mitigate these challenges.

By carefully selecting the appropriate modem preset, you can optimize your Meshtastic network to meet specific communication needs, balancing speed, range, and reliability.

11: PRIVACY AND SECURITY

Ensuring privacy and security is a critical aspect of deploying and managing a Meshtastic® network, especially in scenarios where sensitive communication is involved. Meshtastic uses a combination of encryption, authentication, and secure transmission techniques to safeguard data. Its security features are designed to protect against eavesdropping, unauthorized access, and data tampering.

Here are some of the security features offered by Meshtastic:

- **End-to-End Encryption:** Messages are encrypted on the sender's device and decrypted only on the recipient's device.
- **The encryption standard** is AES-256-GCM (Advanced Encryption Standard with Galois/Counter Mode).
- **Channel Keys:** Each Meshtastic channel is secured with a unique encryption key. Only devices configured with the correct key can communicate.
- **Decentralized Network:** Meshtastic networks operate without central servers, reducing the risk of single points of failure or centralized data breaches.

CONFIGURING CHANNEL SECURITY | To ensure secure messaging, each Meshtastic channel has its own encryption key. Only devices with the correct key can communicate to each other on a particular channel. The default LongFast channel uses a simple four-character encryption key (AQ = =) that allows all users to access it. If security is not a concern, you can leave this default setting. But if you want to keep a channel with the name LongFast and limit access to only trusted parties, you can create a new, secure key for that channel. To do this, go to the Radio Configuration page and open the Channels tab. Select the LongFast channel. In the PSK dialog box, tap the circular arrow to generate a new key, then press the save button, and send the settings to the device. It will automatically reboot. The secure channel can now be shared with trusted parties.

TELEMETRY | Meshtastic devices periodically send data packets to the network to report current readings from sensors. Battery voltage, channel utilization, and for some devices, temperature, and other weather readings are transmitted. If this information is not needed for your use case, consider adjusting the settings accordingly. To do so, navigate to the Radio Configuration page and open the Telemetry tab. The default interval for most settings is 900 seconds (15 minutes.) You can set the intervals for once every two hours (7,200 seconds) or whatever value suits your needs. Longer intervals will conserve power, reduce network congestion, and provide fewer opportunities for your nodes to be discovered by others.

POSITION | Another security concern is the feature that allows the device's physical location to be sent to the network. GPS derived or static position reports are sent over the primary channel. In most cases, this is the LongFast channel. Disabling position reporting is critical to ensuring a secure operating environment, and there

11: PRIVACY AND SECURITY

is an option to disable this function. To do so, navigate to the Radio Configuration page. In the Channels tab, open the primary channel, which is at the top of the list. (It will be numbered 0.) In the dialog box, toggle off the "Position enabled" option. Save this setting, then press send. The device will reboot. Additionally, you can adjust or disable the transmission of all location data in the Position tab of the Radio Configuration page.

Another option is to clone the LongFast channel and use the primary (zero) channel for private communication by assigning it a unique encryption key. In this setup, there are two LongFast channels. One has a private encryption key and is the primary channel. The second LongFast channel uses the default (AQ = =) key and can communicate with the public. To do this, navigate to the Radio Configuration page and open the Channels tab. Create a new channel. Assign it the name LongFast. In the PSK box enter AQ = = and save this setting. Open the other LongFast channel. (The one numbered 0.) Generate a new encryption key. Save this setting and send the data to the device, which will reboot.

SECURE PIN | If your device does not have a display screen, it will use the default PIN 123456 for pairing to your phone. This allows virtually anyone to pair to your device. For better security, set a custom PIN that only you know. To do this, navigate to the Radio Configuration page. Under the "Bluetooth" tab, the "Pairing mode" function has options for using a random PIN or a fixed PIN. Devices with a screen can display a randomly generated PIN, but those without a screen must use a fixed PIN. Select the option "fixed PIN" from the dropdown menu and enter a PIN you can remember in the dialog box. Press save and send the settings to the device, which will reboot.

LOST OR STOLEN DEVICE | Your network is secure as long as you control all the devices connected to it. If you've built a secure network and someone steals a device or if one becomes lost, it compromises the security of the entire network. If this happens, immediately generate and share new encryption keys for the channels used by the remaining nodes on the network.

12: ANTENNA CONSIDERATIONS

The two most important components in a radio communication system are the radio itself and the antenna. Some would argue that the quality of the antenna is more important than the quality of the radio. A poor-quality radio will often perform well with an excellent antenna while a high-performance radio with a low-quality antenna will perform poorly.

The performance of a Meshtastic® network depends heavily on the quality and tuning of its antennas. A poorly tuned antenna can significantly decrease the range, signal quality, and reliability of your network. In this chapter, we'll discuss common problems caused by poorly tuned antennas and provide practical tips and solutions to improve their performance.

Two factors should be considered when evaluating antennas: how well tuned the antenna is for the frequency on which the radio will transmit, and the efficiency of the antenna as measured by standing wave ratio.

Standing wave ratio (SWR) is a comparison of the power radiated from an antenna with the power reflected back toward the radio. This value is expressed as a ratio. A ratio of 1:1 is considered perfect. An SWR of 2:1 is less than ideal, but acceptable. A value of 3:1 is usable, but inefficient. If the SWR of an antenna is above

4:1, it may damage the radio. I never deploy an antenna without measuring its performance with a vector network analyzer (VNA). The Nano VNA is an inexpensive tool for this job. There are many good YouTube videos that explain how to calibrate and use it.

The other factor to consider is how well tuned an antenna is to the frequency being used. In the U.S., Meshtastic uses frequencies between 902 and 928 MHz, with the center frequency being 915 MHz. A Meshtastic antenna is basically a wire cut to a certain length. The length of the wire determines the frequency at which a radio wave will be transmitted on it effectively. This property is called *resonance*. Short wires are resonant at high frequencies. Long wires are resonant at low frequencies. The goal for our purposes is to find an antenna that is resonant on the frequency of 915 MHz.

Heltec sells a LoRa board for use with Meshtastic that comes with a small wire antenna (pictured below). It comes with an optional plastic shell that protects the components. The supplied antenna, though it is not elegantly designed, is well tuned for 915 MHz.

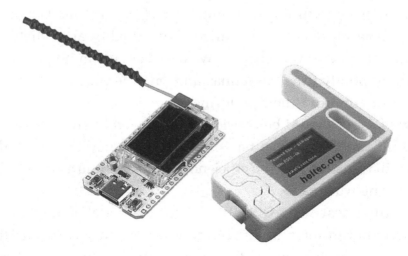

More often, a different antenna arrangement is needed for Meshtastic. Most durable cases that enclose Meshtastic nodes have

12: ANTENNA CONSIDERATIONS

a ¼ inch hole that serves as a port for an antenna with an SMA type connector. An antenna with a threaded SMA connector is attached to a short cable, and the cable is connected to the LoRa board. Most LoRa boards use IPEX connectors, as shown in the diagram below.

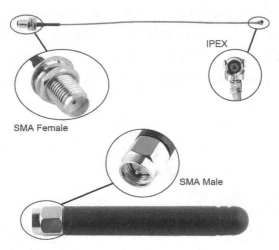

Note in the diagram, there are both male and female SMA connectors. The male version has a pin protruding from the center. The female version has a receptacle that accepts the pin. When buying an antenna, make sure you have the proper mating connector style.

I purchased a batch of ten inexpensive LoRa antennas that were advertised for the 915 MHz frequency. Before attempting to use one, I measured its performance. Although the antenna measured a very low SWR of 1:1.2, that SWR was found at 790 MHZ. The SWR at 915 MHz was greater than 5:1. And it wasn't just one antenna. I tested three more from the same batch and they all showed the same results. Using them for Meshtastic would not just result in poor signal transmission, but these antennas could damage the radio.

With few exceptions, the stubby antennas provided with most LoRa boards offer mediocre performance. The SMA antennas included with Rak Wireless and Heltec boards have SWRs between 2.3:1 and 2.8:1 at 915 MHz. They won't damage a radio, but they won't perform well either. I can recommend a few antennas that offer better performance.

Rak Wireless sells an antenna called the "Blade." There are many antennas that resemble it, but it can be distinguished by its wedge-shaped tip at the end of the vertical portion. My measurements show SWRs between 1.3:1 and 1.8:1. These units cost $6 on the Rak Wireless website. The Blade has an articulating joint.

For nodes used in elevated locations, you might consider an antenna sold on Amazon and Ali Express. This antenna is advertised as having 12dbi of gain. My measurements show SWRs on 915 MHz between 1.3:1 and 1.8:1. They cost between $2 and $10 depending on where they are purchased. I can't verify the claim of 12dBi of gain, but I've had good success using them with nodes deployed on my roof and on mountaintops. This antenna has an articulating joint.

DIRECTIONAL ANTENNAS | The antennas mentioned above disperse radio signals in all directions when positioned vertically. Thus, they are called omnidirectional. When you don't know in which

12: ANTENNA CONSIDERATIONS

direction a listening station might be, it's best to use an antenna that covers 360 degrees. The benefit of these antennas is that some signal has a chance to be sent in the direction of an unknown receiving station. The drawback with *omnidirectional* antennas is that most of the signal is never heard by a listening station, and it is wasted. If you know where a receiving station is located, it would be more desirable to send a signal only in that direction. When a radio signal is sent in a single direction, less of it is wasted and the signal is stronger. A *directional* antenna can be used with Meshtastic radios to send a stronger signal in a particular direction. A Yagi antenna (pictured below) is one such antenna. It consists of a beam with several cross elements arranged at 90 degrees to the beam. Directional antennas offer higher gain and are useful when a stronger signal is desired and the listening station is in a known location.

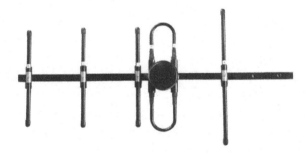

A wide variety of high gain antennas are available on websites that specialize in LoRa equipment. Careful consideration should be given when buying equipment for Meshtastic. Be sure the antenna you purchase is tuned to the correct frequency. When connecting a cable from a radio to an antenna, realize that the longer the cable is, the more signal loss you will experience. Ham radio operators are accustomed to using a radio at ground level with a long coaxial cable connected to an elevated antenna. This arrangement may not work for Meshtastic. More often, the radio itself is deployed at elevation, but there may be ways to engineer a compromise. Be sure to have the right type of connectors to build your system. For nodes deployed outdoors, use weatherproofing sealant, heat shrink tubing, or rubber tape to keep water out of joints.

13: ELEVATED LOCATIONS

When Meshtastic® devices are positioned correctly, one may build a network that covers a city, a county, or even a state. In this chapter, we'll learn how to deploy elevated nodes that form the backbone of a wide area network.

LoRa radios operate in the UHF radio band. UHF radios transmit signals that can be received by other radios providing there is "line of sight" between the sending and receiving stations. Due to the curvature of the earth, at sea level, the direct line of sight for two radios is approximately three miles. This is one reason why UHF radios have a short operating range. The distance over which a radio can successfully transmit a signal is called the *radio horizon*.

A second problem for UHF radio communication is absorption of signals by vegetation and buildings. The more trees and buildings there are between sending and receiving stations, the shorter the distance will be where signals can get through. Placing a radio in an elevated location can remove buildings and vegetation from the signal path and it extends the radio horizon.

Deploying a Meshtastic node in an elevated location is an excellent way to extend the range of your mesh network and provide reliable communication in hard-to-reach areas.

Here's a detailed guide covering strategies, tactics, and considerations for deploying such nodes:

Radio waves in the lower UHF band *can* penetrate buildings and vegetation to some degree. However, radio frequencies between 2 and 5 Gigahertz are easily blocked by vegetation. Signals in the 900 MHz band lie somewhere in the middle, and may overcome a blocked signal path.

I have several Meshtastic nodes deployed at a radio site at Usery Pass in Mesa, Arizona. The radio site is situated in a saddle with peaks several hundred feet high to the north and south. My home does not have a clear line of sight to my nodes at Usery Pass. Nevertheless, even the weak signals of my LoRa device get through to my rooftop node directly, without needing an intermediate node to relay them. When planning your network, consider deploying a node in an elevated location with less-than-ideal line of sight. It may prove to be more useful for transmitting signals than you imagined. Testing will help determine the best deployment locations.

It's possible to build a Meshtastic network that covers hundreds of miles. The secret is deploying nodes in strategic locations. Usery Pass at 2,300 feet of elevation, lies to the east of Phoenix, Arizona. The White Tank Mountains at 4,000 feet, lie to the west. When nodes were deployed at these locations, it created a backbone for a citywide network. When a node was deployed on Pinal Peak, 50 miles to the east of Phoenix, at an elevation of 5,000 feet, it expanded the coverage of the Maricopa County network to include Pinal County. When a node was deployed on Mt. Lemmon, at 9,000 feet, 100 miles to the south, it expanded the network to Pima County. When a node was deployed on Greens Peak at 10,000 feet, it connected eastern, southern, central, and northern Arizona into a single network spanning hundreds of miles.

In this case, the network is more a proof of concept than a functional communication pathway. What follows is my

13: ELEVATED LOCATIONS

non-expert assessment of the problem we encountered and what I think may be the solution.

The cities of Phoenix and Tucson have hundreds (perhaps thousands) of Meshtastic nodes. Many of them are unable to reliably send direct messages. I suspect the problem is that nearly all the traffic is passed on the same default channel (frequency slot 20 in the U.S.) When hundreds of nodes simultaneously send signals on a narrow radio spectrum, many of the signals won't reach their intended destination due to excessive noise. For a far-reaching network to function properly, congested urban areas would probably need to build subnetworks that utilize an alternate frequency slot.

Breaking a larger network into subnets would solve the problem of excessive signals clogging up a single frequency. The goal would be to group small communities into zones on separate frequencies within the metropolitan area. If desired, one frequency could be dedicated to emergency traffic of regional or statewide interest. The nodes for the emergency network could be deployed in strategic, elevated locations throughout the region.

The other problem is that most nodes in Arizona use the default Long Range/Fast modem setting. While this preset passes data at an adequate speed for a neighborhood network, it is far too slow for one with hundreds of nodes. For messages to be sent reliably in a large network, a modem preset should be chosen that allows faster data transmission.

The Arizona Mesh Organization is experimenting with this concept. We're building a network across Maricopa County with some nodes at elevated locations and others at ground level. We're utilizing an alternate frequency slot to avoid the congestion found on the default frequency slot 20. So far, we've had decent success passing direct messages and posting on channels. We're monitoring our ability to reliably send messages and plan to test different modem presets to optimize efficiency and reliability.

Power consumption should be considered when planning the deployment of elevated nodes. One way to reduce power consumption is to activate "Power Saving" mode. When activated, this feature disables Bluetooth, serial, and Wi-Fi, and the device's display screen is disabled. This is particularly beneficial for devices relying on low-energy power sources such as a small solar panel and battery. If power saving mode is enabled, modifications to settings can be made by waking the device through pressing the user button, resetting, or the feature that allows remote administration. To activate power saving mode, go to the Radio Configurations page and open the "Power" tab. Toggle the switch on that says "Enable power saving mode."

For ESP32 boards, the Router role enables Power Saving mode by default. Because Bluetooth is disabled in this mode, special consideration must be given when updating settings. When you need to modify the settings of such a node by Bluetooth, it's recommended that you use remote administration to temporarily change the role of the node to Client in order to make the update, and then revert to Router when you are done. Remote node administration is covered in the next chapter.

Nodes deployed at elevated locations must offer protection from the elements, and ideally, be solar powered. One such option is the RAK Wireless Solar Unify Enclosure Case sold on the Rokland website. This enclosure is rated at IP67 for water ingress. The Rokland website has several versions of this enclosure. The smaller ones do not generate enough current through the solar panels to keep a Meshtastic node operational. The model I recommend measures 150x100x45mm. It's an injection molded plastic enclosure with two halves that fit together like a clamshell with a rubber gasket between the front and back halves to keep moisture out. The front of the unit has a built-in solar panel. The back has recessed threaded brass inserts for mounting on a pole or other structure. Rokland sells a kit for mounting this unit to

13: ELEVATED LOCATIONS

a pole. In addition to the enclosure, you would need to purchase a rechargeable battery and a Rak 4631 board. (A Meshtastic starter kit is available on their website.) Be advised: this enclosure uses a reverse polarity antenna, which is included. If you plan to use a different antenna, be sure it has reverse polarity, or if not, use an appropriate adapter. Also consider that the battery connector for the Rak 4631 board accepts a 2.0mm JST connector, so buy a battery with a compatible connector.

Assembly of this unit is straightforward. A supplied mounting board is screwed to the inside of the enclosure and the Rak board is secured to it using 4 screws. Bluetooth and LoRa antennas are connected and installed. A battery is connected and secured to the mounting board with 3M double sided foam tape. The solar panel is connected, then the rubber gasket is positioned. Lastly, the front and back halves are fastened together with screws through holes in the back. I've deployed several of these nodes in remote locations with good results. The Rak 4631 board used in this device when paired with a 3,000 mAh lithium-ion battery can run for 10 days on a single charge.

I designed a mount that bolts to the rear of the solar enclosure to fit on a vertical mast and printed it with a 3D printer. One mount is fastened to the upper portion of the rear of the enclosure and a second one is secured to the lower portion. The unit is then secured to a vertical mast using a hose clamp, which passes through an open channel in the mount.

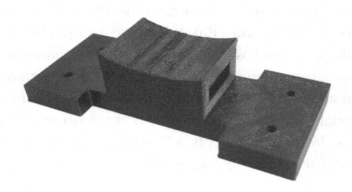

Another option for elevated deployment is building your own solar powered device. I've built dozens of these units and they've been deployed around the state of Arizona during the last two years. Many people have engineered solar powered units for outdoor use, but I have yet to see one that exceeds this design for durability and efficiency.

The design I will explain is based on a model I purchased from a vendor on Etsy. I modified the original design to suit my needs and I offer it here as a suggestion of what is possible.

Below is an image of the parts:

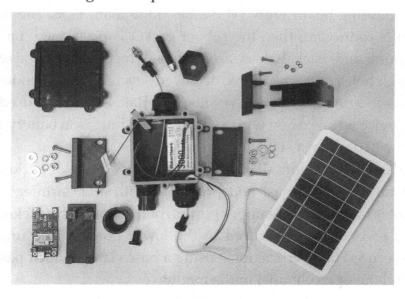

13: ELEVATED LOCATIONS

What follows is a description of how the parts are assembled.

This node is built using a waterproof three port electrical junction box that sells on Amazon for less than $10 (shown in the center of the image). The 5-volt, 2-watt solar panel used in this application is available on Ali Express for $2. The 3,000 mAh battery costs $7 on Amazon. The Rak 4631 Meshtastic starter kit is $36 on Amazon. In the upper right corner of the image are a couple of 3D printed parts that bolt together. The smaller part is adhered to the back of the solar panel with weatherproof 3M double-sided tape. The two parts, when bolted together, create a mount that can be secured to a mast with a hose clamp or screwed into wood.

In the lower right of the image, the solar panel's cord has its USB connector cut off. It passes through a 3D printed dowel with a 0.14 diameter hole through it. The solar panel cord is spliced to a 1.25 mm JST connector, which attaches to the Rak Board's solar panel socket. The passageway in the dowel through which the solar panel cord passes is sealed with silicone sealant.

The 3.7 volt, 3,000 mAh MakerHawk battery has its 1.25 mm JST connector removed and a 2.0 mm JST connector is spliced on. (Similar batteries are available with a 2.0 mm JST connector if you'd rather not bother with removing and attaching the correct connector.) After the connector has been spliced on, it is connected to the battery receptacle on the Rak board and secured inside the enclosure with double sided tape.

On the left side of the junction box is a thin Bluetooth antenna cable with an IPEX connector that is secured to the BLE socket on the Rak board.

In the lower left of the image, a 3D printed base plate is glued to the bottom of the enclosure beside the battery. It has 4 square, raised posts to allow room for the components that protrude from the bottom of the Rak board. (The base plate and Rak board are in the correct orientation for assembly.)

Pilot holes are drilled in the base plate and the Rak board is screwed to it using 4 screws in the corner holes. The board is aligned inside the enclosure so the USB connector is accessible through the round port in the junction box that is to the lower left in the image. This port (which has the cap removed) is fitted with a solid, round 3D printed 0.30 diameter dowel to keep dust and moisture out.

In the upper center, there is an IPEX cable that connects to the LoRa socket on the Rak board. It passes through the uppermost round opening of the enclosure and then through a 3D printed dowel that has a 0.18 diameter hole through it. After fitting the antenna cable inside the dowel, the dowel, where it meets the base of the threaded antenna connector, is sealed with silicone sealant.

A hexagonal 3D printed cap with a ¼ inch hole through the center is fitted over the top of the threaded cap and the antenna connector protrudes through it. Silicone sealant is applied to the top face of the dowel before the cap is installed. A nut is secured on the antenna post. Then the antenna is screwed onto the threaded SMA post.

Lastly, to either side of the junction box, there are 3D printed L-shaped brackets. These are bolted to the back of the junction box. The vertical portion of each bracket has a slot through which a hose clamp can be passed for securing the device to a mast.

The images that follow show both mounting options for the solar powered unit. This image depicts the solar powered device mounted to a vertical mast.

The hose clamp passes through the openings in the L-shaped brackets that are bolted to the back of the node. The hose clamp

13: ELEVATED LOCATIONS

then passes through a groove in the arm of the solar panel mount. **Warning:** 3D printed parts are easily crushed when clamps are overtightened. These mounts are designed so that the clamp should be tightened just until it is snug and no more.

Note that the radio is positioned at the top of the mast with the antenna well above it. When a radio antenna is deployed alongside a metal vertical mast, the mast interferes with the transmission of the signal. The best practice is to deploy the node at the top of the mast, so the antenna has no metal objects next to it.

The image below shows this unit deployed at rooftop elevation.

Both components are secured to a wood beam with screws. In this image, the top of the antenna is not visible, but it extends above the roofline. (This is one of the oldest units I own. It's been in service for two years. Note that the hexagonal cap for the lower lefthand port is missing. I used rubber tape to waterproof the opening and vinyl tape was applied over it for UV protection.)

Many amateur radio clubs have access to towers at elevated locations where Meshtastic devices could be deployed. No power source is required, no rack space, and no internet cables are needed. Alternatively, a solar powered, weatherproof node can easily be attached to the branch of a tree on a prominent hill or mountain and camouflaged. Some nodes are currently deployed in Phoenix on church steeples. The possibilities are endless. When modem settings and frequency slots are optimized, one can build a statewide network with only a few radios positioned in strategic locations.

14: REMOTE NODE ADMINISTRATION

Nodes that are placed in elevated and remote locations may occasionally require settings to be updated. Because these devices can be difficult to access in person, the Meshtastic® development team has provided a way to update the settings remotely. Remote administration is a handy tool when you need to manage a device without relying on Bluetooth or Wi-Fi connections after the initial setup is complete. Remote node administration previously required setting up an admin channel. The method used on newer firmware versions is much easier and it is the one I recommend.

For users of firmware version 2.5 or higher, follow these steps to set up remote administration:

- On the local device that will be used to change settings on the remote node, connect your phone to the device.

- In the mobile app, navigate to the Radio Configuration page, and open the Security tab. Note that there is a public and a private key. Copy and save the public key by pressing on the text to highlight it, select the entire text and then select "copy."

- Navigate to the Connection tab and disconnect from the admin node then connect to the node you want to control remotely.

- Navigate to the Radio Configuration page and open the Security tab. Under Admin Key press the "Add" button. Tap on the dialog box that says 1/3. In this field, paste the public key that was copied from the admin device by pressing in the text field.

- Up to 3 admin keys may be added, one per field, allowing up to 3 nodes to remotely administer this node.

- When you are finished, press Send. The device will reboot.

To verify that the admin node can change the settings of the remote node, open the Meshtastic app and connect to the local node you will use to control the remote node.

- Navigate to the device list. Tap on the oval with the four-character name of the device that is to be configured remotely.

- From the drop-down menu, select Radio Configuration.

- You can navigate through all the configuration settings *from* the admin node and change them as you please *on* the remote device, just as if you were connected to the remote device by Bluetooth. Data is sent through the mesh network. Try changing the name of the remote node to get a feel for how this feature operates. When you are through, tap the back arrow to navigate through the previously viewed screens to exit remote administration mode.

15: MESHTASTIC AND THE INTERNET

Meshtastic® creates communication networks that are independent of central servers and do not require an internet connection. Nevertheless, LoRa networks can be connected to the internet using an MQTT server. MQTT stands for Message Queuing Telemetry Transport. Here's how it works in simple terms for Meshtastic nodes:

- Think of MQTT as something like a postal service for messages in a digital world where Meshtastic devices are the houses.

- In this postal service, there's a central post office called the "broker". When a Meshtastic device wants to send or receive a particular type of message, it talks to the broker instead of to other nodes. The broker helps manage all the communication, making sure messages go where they need to go.

- MQTT uses two features called *subscribing* and *publishing*.

- Publishing: When one of your devices has a message (like

"I've detected smoke!" from a smoke detector), it sends this message to the broker. This is called publishing.

- Subscribing: Other devices that want to receive this message tell the broker they're interested. This is subscribing. For example, your phone might subscribe to updates from your home security system.

- Messages are sorted by topics. Just like mail could be sorted by different types (like bills, letters, packages), in MQTT, messages are tagged with topics. If a device publishes a message to the topic "home/smoke", only the devices that have subscribed to this topic will get that message.

- Since the devices only need to connect to the broker and not to each other directly, it saves energy and bandwidth. It's especially useful in places where direct connections might be hard or where devices might go in and out of range.

- In a Meshtastic network, each device can act like a small post office (meshing messages between each other), but when it comes to sending messages outside the local network or to a server for logging or control, MQTT helps. Your Meshtastic device might connect to a broker on the internet, allowing your message to reach devices or services that aren't available in your immediate mesh network.

MQTT helps your Meshtastic devices communicate efficiently by using a central point (the broker) to handle the distribution of messages, making it easier for all devices to stay in sync or share information without needing direct connections between each pair.

15: MESHTASTIC AND THE INTERNET

There are several disadvantages to using MQTT with Meshtastic that might affect its functionality or efficiency:

PRIVACY | When messages are sent through a public MQTT broker, there's a risk of privacy invasion if those messages aren't encrypted or if they're shared more widely than intended. Recent changes by Meshtastic to limit topic subscriptions aim to mitigate this, but it's still a potential issue, especially with location data.

SECURITY | If not configured properly, MQTT can be vulnerable to security threats. While MQTT can use encryption, setting it up correctly to ensure all communications are secure requires some technical know-how. If encryption isn't enabled, messages could be intercepted or read by unauthorized parties.

NETWORK DEPENDENCY | MQTT relies on an internet connection for the broker. If your Meshtastic device can't connect to the internet (like in a remote area without service), then the MQTT part of your communication won't work, limiting you to local mesh networking.

POWER DEMAND | For battery-operated devices, constantly connecting to an MQTT broker can drain battery life. Since every message needs to go through the broker, this won't be as power-efficient as direct mesh communication for devices that run on batteries for extended periods. This is particularly true if the device needs to reconnect frequently.

TRAFFIC AND SERVER LOAD | The Meshtastic public MQTT server might experience high traffic, leading to potential overloads or restrictions on what can be communicated. This can affect the performance of your network, especially if many users are trying to use the same service.

LATENCY | While MQTT is designed for efficiency, adding an internet hop (to the broker and back) can introduce latency compared to direct mesh communication, which could be a drawback in scenarios where immediate response is critical.

CONFIGURATION COMPLEXITY | Setting up your own MQTT server or correctly configuring a device to interact with either a public or private MQTT server can be complex for users not familiar with networking concepts. This setup might deter less tech-savvy individuals from fully utilizing this feature.

INFORMATION SHARING CONTROL | With MQTT, controlling who gets what information can be tricky. While the broker manages subscriptions, ensuring that only intended recipients get messages requires careful setup or understanding of topic hierarchies, which might not be intuitive for all users.

By understanding the advantages and disadvantages, users can better decide if MQTT is suitable for their Meshtastic setup or if they need to take additional measures like using private servers, ensuring proper encryption, or considering alternative protocols for some applications.

16: MESHTASTIC AND AMATEUR RADIO

One of the bands used by licensed amateur radio operators (hams) overlaps with the U.S. LoRa radio band of 902-928 MHz. While non-licensed Meshtastic® use is limited to 1 watt ERP, amateur radio is allowed up to 10 watts PEP. Hams can use power amplifiers and high gain antennas to boost their signal and thereby increase coverage area. But there are additional tradeoffs for amateur use.

When using Meshtastic as an amateur radio operator, a toggle switch is flipped on the mobile app, which disables the use of encryption. One of the most valuable features of Meshtastic is not available to hams because encryption could be used to obscure the meaning of a message which is not allowed under Federal Communication Commission (FCC) regulations. Amateurs are also required to identify themselves periodically with their call sign. Thus, privacy and security are compromised.

The loss of encryption and privacy may discourage the use of Meshtastic under these terms. You can use Meshtastic as a ham, but who are you going to communicate with? Ham radio operators make up a large segment of Meshtastic users, but many are using the unlicensed option because it connects them to ordinary people, allows the use of encryption, and has the option of anonymity.

All users should be aware that their unlicensed Meshtastic device is governed by FCC Part 15 Rules & Regulations. We've all seen the labels that come with electronic products that say, "This device must accept any interference and must not produce any interference..."

Here are some major points our Meshtastic device meets for unlicensed operation in the USA:

- Part 15.247 (a)(2) -- In regards to digital modulation techniques (i.e., using LoRa in the 902-928 MHz band);

- Part 15.247 (b)(3)(4) -- An effective radiated power (ERP) of 1W (+30dBm) with reduced power if antenna gain is greater than 6 dBi;

- Part 15.247 (d) -- With proper signal quality (or cleanliness);

- Part 15.247 (e) -- And meeting power spectral density requirements (e.g., Meshtastic's eight LoRa radio presets).

The above Rules & Regulations (FCC website, fcc.gov) are located in Title 47 of the Code of Federal Regulations (CFR) often referred to as "47 CFR Part 15" or just "FCC Part 15" in casual usage.

There is value in the amateur Meshtastic space, especially for those who want to experiment (or as described by the FCC as "... contribute to the advancement of the radio art). The lower UHF frequencies have better propagation through foliage and buildings than UHF frequencies available for spread spectrum LoRa technology. With a LoRa radio that offers multiple bands, one could test the ability of signals on the 420 MHz amateur band to reach locations that are out of the question for 900 MHz, especially if one incorporated a power amplifier and high gain antenna.

16: MESHTASTIC AND AMATEUR RADIO

Here are some FCC Part 97 Rules & Regulations a ham should read and consider when placing a Meshtastic device into the amateur radio service:

- Part 97.303 Frequency sharing requirements.

    - As they normally apply to amateur stations, including ISM band usage and geographical restrictions.

    - Part 97.305 Authorized emissions types.

    - Spread Spectrum (SS) is allowed in the UHF bands.

    - 70 cm band - Although SS digital coding is allowed here, it's not practical due to its bandwidth limitation of 100 KHz, e.g., limited to Meshtastic's slowest preset.

    - 33 cm band - This is the primary band for using our devices in the USA, also known as the 915 ISM band. Higher frequencies are possible but no Meshtastic devices are currently available.

- Part 97.307 Emission standards.

    - The amateur operator is expected to follow good technical practice regarding bandwidth used and any limitations that may apply.

- Part 97.309 RTTY and data emission codes.

    - An acceptable data emission is one using a digital code that is documented publicly (such as the LoRa protocol by Semtech and digital encoding by Meshtastic).

- Part 97.311 SS emission types.

    - The amateur operator is expected to follow good operating procedures, i.e., the SS emission itself is not to be used to obscure or hide communication.

- Part 97.313 Transmitter power standards.

    - The SS transmitter output will not exceed 10 W PEP. There is no mention of antenna gain restriction. Follow appropriate RF exposure guidelines and FCC limitations.

- Part 97.3 Definitions.

    - ERP (Effective Radiated Power) - The product of the power supplied to the antenna and its gain relative to a half-wave dipole in a given direction.

    - PEP (Peak Envelope Power) - The average power that is supplied to the antenna transmission line by a transmitter.

    - SS (Spread Spectrum) - Emission type using bandwidth-expansion modulation with designators A, C, D, F, G, H, J or R as the first symbol; X as the second symbol; X as the third symbol. This is essentially an amplitude or angle-modulated main carrier for digital data transmission (such as LoRa). Example designators include F1D, G1D or F/G1X. Note, designators are described in FCC Part 2.201 Emission, modulation and transmission characteristics.

16: MESHTASTIC AND AMATEUR RADIO

Here are two examples of excellent amateur radio club websites that talk about their use of Meshtastic and some of the apps and tools that are available:

- http://www.nemarc.org/mesh.html

- https://kc4rc.com/meshtastic/

It is unknown if amateur use of Meshtastic® will grow in popularity, but some users will take advantage of the unique opportunities it provides. There are members of my local ham radio club who are doing this right now. If you decide to go this route, be sure to purchase a LoRa board that transmits on the frequencies you intend to use. Buy antennas that are matched to that frequency and batteries that are appropriate for your use case. For the ham, don't stop here; keep experimenting; keep moving forward!

ANOTHER BOOK YOU MAY ENJOY

EMERGENCY PREPAREDNESS AND OFF-GRID COMMUNICATION

This book discusses storing food and water, securing your home, developing a bug-out plan, emergency first aid, personal hygiene during a crisis, backup power options, home security, self-defense, surviving nuclear war, and protecting your family against the worst consequences of an economic collapse.

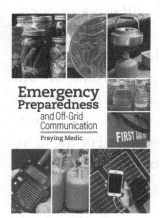

On top of all those crucial topics, it is packed with information about emergency communication. It contains details in more than a dozen chapters that explain the options for communicating when cellular and internet services are unavailable and the power grid is down. (Available at most online book retailers.)